I0468617

HUE
Coloring

50 WILD TIGER DESIGNS
An Adult Coloring Book

ISBN-13: 978-1530888771
ISBN-10:153088871
Copyright © 2023 Hue Coloring
All rights reserved.

No part of this publication may be copied, reproduced in any format, by any means, electronic or otherwise, without prior consent from the copyright owner and publisher of this book.

IN THIS COLORING BOOK...

50 Wild Tiger Designs are included in this adult coloring book to help you relax and make your life more colorful. These illustrations are created for you to bring enjoyment to life, and designed with beautiful patterns that appeal to adult eyes.

TIPS TO A RELAXING COLORING

Find a quiet space. It's easier to focus on what you are doing when there are no distractions.

Organize your materials. Lay out your coloring book and crayons, pens, or pencils.

Set the mood. Turn on some tranquil music, diffuse lavender or another relaxing oil, and make sure you have your preferred drink at hand.

Select your picture. Which image speaks to you today? That's the one you should color. Choose your pallette. Select the colors you will be using for your image.

Begin coloring. This is the fun part. Don't worry about getting evrything perfect' just start. If you feel you don't want to do it anymore, just stop!

SHOWS US YOUR CREATION!

We'd love to hear from you, show us what you created.
Facebook: www.facebook.com/huecoloring
Email: huecoloringbooks@gmail.com

www.ingramcontent.com/pod-product-compliance
Lightning Source LLC
Chambersburg PA
CBHW080711190526

45169CB00006B/2334